ROSEN Verified
NATURAL DISASTERS

DROUGHTS

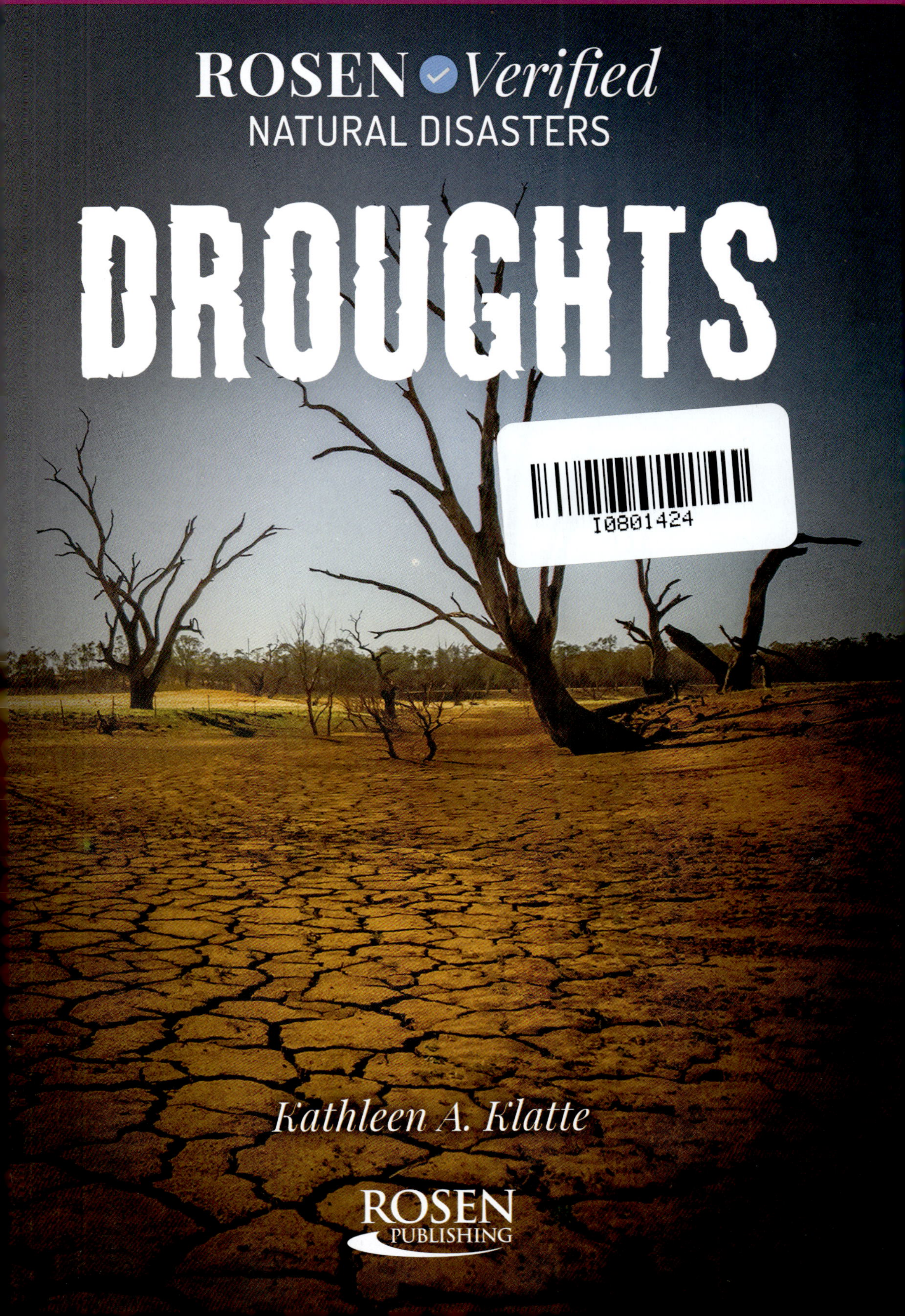

Kathleen A. Klatte

ROSEN PUBLISHING

Published in 2023 by The Rosen Publishing Group, Inc.
2544 Clinton Street, Buffalo, NY 14224

First Edition

Editor: Greg Roza
Book Design: Michael Flynn

Photo Credits: Cover idiz/Shutterstock.com; (series background) PinkPueblo/Shutterstock.com; p. 5 R_Tee/Shutterstock.com; p. 7 Jon Manjeot/Shutterstock.com; p. 8 Milju varghese/Shutterstock.com; p. 9 Checco2/Shutterstock.com; p. 11 Simone Hogan/Shutterstock.com; p. 13 Chad Robertson/Shutterstock.com; p. 15 Douglas Cliff/Shutterstock.com; p. 17 VectorMine/Shutterstock.com; p. 18 monticello/Shutterstock.com; p. 19 Tanja Esser/Shutterstock.com; p. 21 zmpikes/Shutterstock.com; p. 23 Nelson Antoine/Shutterstock.com; p. 25 Everett Collection/Shutterstock.com; p. 27 (Geldof) s_bukley/Shutterstock.com; p. 27 (Live Aid) https://commons.wikimedia.org/wiki/File:Live_Aid_at_JFK_Stadium,_Philadelphia,_PA.jpg; p. 29 Kaikoro/Shutterstock.com; p. 31 AP Photo/Rich Pedroncelli; p. 33 Simone Hogan/Shutterstock.com; p.35 Antonello Marangi/Shutterstock.com; p. 36 llaszlo/Shutterstock.com; p. 37 https://commons.wikimedia.org/wiki/File:Pressler_drill_5_beentree.jpg; p. 39 Joao Virissimo/Shutterstock.com; p. 40 ShutterLibrary/Shutterstock.com; p. 41 Joseph Sohm/Shutterstock.com; p. 43 Tatiana Grozetskaya/Shutterstock.com; p. 45 STEKLO/Shutterstock.com.

Library of Congress Cataloging-in-Publication Data

Names: Klatte, Kathleen A., author.
Title: Droughts / Kathleen A. Klatte.
Description: Buffalo, New York : Rosen Publishing, [2023] | Series: Rosen verified: natural disasters | Includes index.
Identifiers: LCCN 2022023372 (print) | LCCN 2022023373 (ebook) | ISBN 9781499469462 (library binding) | ISBN 9781499469455 (paperback) | ISBN 9781499469479 (ebook)
Subjects: LCSH: Droughts—Juvenile literature.
Classification: LCC QC929.25 .K53 2023 (print) | LCC QC929.25 (ebook) | DDC 551.57/73—dc23/eng/20220524
LC record available at https://lccn.loc.gov/2022023372
LC ebook record available at https://lccn.loc.gov/2022023373

Manufactured in the United States of America

Some of the images in this book illustrate individuals who are models. The depictions do not imply actual situations or events.

CPSIA Compliance Information: Batch #CWRYA23. For further information, contact Rosen Publishing at 1-800-237-9932.

CONTENTS

RAIN, RAIN, GO AWAY!

Rainy days can be a pain. They can make the ground muddy and mess up your hair. Plans might be canceled. You have to remember your umbrella. You might even need a jacket or boots. That's no fun!

Earth needs water to work right. Plants need water to grow. People and animals need water to drink. Earth gets that water from rain. When there isn't enough rain for a long time, it's called a drought.

Droughts can happen all over the world. Some places are naturally very dry all the time. Others have a rainy season and a dry season each year. Scientists recognize four kinds of drought: permanent, seasonal, unpredictable, and invisible.

Access to water is a basic human right. It's recognized by the United Nations (UN). Everyone needs clean water for drinking, cooking, and bathing.

Droughts can cause crops to die. They can damage Earth in ways that take a long time to heal. Changes in wind or ocean currents cause droughts. Droughts can also be caused by things people do.

Luckily, there are also things people can do to lessen the effects of droughts. Following **conservation** instructions posted by your local government can help people around you. Living a more **eco-friendly** lifestyle can help lessen the effects of global warming for everyone.

PERMANENT DROUGHTS

Deserts are places that receive less than 10 inches (25 cm) of rain annually. They exist in a state of permanent drought. About 20 percent (one-fifth) of Earth is desert. Deserts can be hot or cold. The ground can be rocky or sandy.

Deserts may appear empty, but there are plants and animals if you look carefully! Things that live and grow in the desert have **adapted** to their **environment**. Some animals live underground and only come out at night.

THE SAHARA DESERT

The Sahara is the largest hot desert in the world. It's about the size of the United States. It's home to more than 260 species, or kinds, of animals. This includes the largest kind of bird in the world, the North African ostrich. It also includes the most poisonous type of scorpion, the deathstalker scorpion.

MYTH BUSTERS

MYTHS	FACTS
All deserts are sandy.	Only about 20 percent of deserts are sandy. Others are rocky or mountainous.
All deserts are hot.	The Gobi Desert in Asia is cold. So are the polar deserts.
The Sahara is the world's largest desert.	The largest desert in the world is the Antarctic Desert. It's about 2 million square miles (5.2 million sq km) bigger than the Sahara.

Cacti are plants that have **evolved** to live in the desert. They store water in their tissues for times when there's no rain. They grow sharp spines to prevent animals from eating them.

SEASONAL DROUGHTS

Tropical wet-dry climates exist in places near the equator. They experience seasonal droughts, because there are distinct rainy and dry seasons. Most of the rain—25 inches (63.5 cm) or more—falls for one to three months in the summer. Farmers plant their crops to take advantage of the rainfall.

A monsoon is a type of very strong storm. Monsoons occur during the rainy season in India and southern Asia. In southern Asia, the rainy season is called monsoon season.

During the dry season, it can be hard for animals to find enough water to drink. They may come closer to human settlements in search of water.

The dry season is winter. It lasts much longer. There can be less than 4 inches (10.2 cm) of rain during the entire dry season. A seasonal drought is a normal weather pattern. People have learned to plan for it.

THE EQUATOR

The equator is the imaginary east-west line that divides Earth in half. It's used as a reference point for navigation and for planning travel routes. The areas closest to the equator are called the tropics.

UNPREDICTABLE DROUGHTS

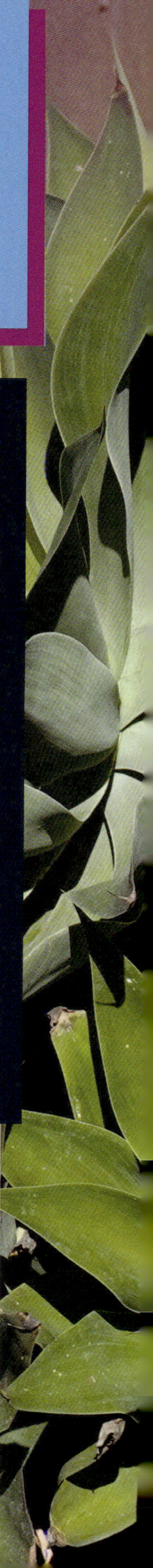

Permanent and seasonal droughts are natural climate patterns. The same amount of rain falls each year. Plants and animals have evolved in those environments. People have adapted to living there.

An unpredictable drought is one that happens in a place that usually receives enough rain on a regular basis. It can be caused by a temporary shift in weather patterns. If less snow and rain fall, there's less water. Human activity can also cause unpredictable droughts.

SCARY NUMBERS!

Currently, almost half of the United States is experiencing drought. This affects:

- 200 million acres (81 million hectares) of crops
- 90 million people
- 30 states

VERIFIED

The Environmental Protection Agency (EPA) is a federal agency. It's responsible for things like water quality. Two important laws it enforces are the Clean Water Act and the Safe Drinking Water Act. You can learn more about the EPA at **www.epa.gov**.

The most important use for clean water is drinking. Plants can be irrigated, or watered, with used water that's been collected and treated. A drip irrigation system uses much less water than spraying plants with a hose.

INVISIBLE DROUGHTS

You've probably lived through an invisible drought and didn't even notice it. Invisible droughts often happen during hot summer weather. Water **evaporates** faster than rainfall can replace it. This type of drought goes away when the seasons change.

Invisible droughts usually aren't dangerous. Local authorities monitor water levels to make sure the drought doesn't get worse. They might ask people to stop watering lawns and washing cars. These activities waste a lot of water.

GARDENING IN A DROUGHT

It's possible to take care of your garden during a drought. The main thing is to use water wisely. Soaking the roots of your plants makes them stronger and saves water. Using a hose carefully is better than letting a sprinkler spray water on the sidewalk.

Using mulch protects the soil, holds in moisture, and keeps pesky weeds from growing!

Opening a fire hydrant on a hot day is dangerous. It wastes a lot of clean water. It also makes it hard for firefighters to do their job.

ALL OVER THE WORLD

Droughts can occur anywhere in the world. In the past, they've occurred on every continent. Parts of Antarctica are in a state of permanent drought. As of 2022, there were droughts in California, Northern Ontario, Central and South America, and Africa.

Droughts are caused by natural factors. Changes in weather patterns might cause less rain and snow to fall. Improper farming and grazing can worsen drought. Native plant species help hold water in the ground. Without them, soil dries out and can blow away.

EL NIÑO

El Niño is a current of warm water that flows along the western coast of South America. It happens every few years, usually around December. El Niño can affect weather patterns all over the world. Some places can have dangerous flooding. Others can have droughts.

The Millennium Drought (2001–2009) was the worst drought in Australian history. Water levels dropped to record lows. The remaining water was too salty to use. Special pipelines had to be built to supply communities with drinking water.

PREDICTING THE FUTURE

Droughts are hard to predict. There are many factors involved. Slight changes in weather patterns can add up over time. If a mild winter is followed by a dry summer, water levels can start to drop.

Unlike other natural disasters, droughts don't show up as radar images. Small factors mix together over time to cause a drought. People need to pay attention when scientists warn that a drought is possible. They should also be mindful of their water **consumption**.

DO YOUR PART

A drought is different from other types of natural disasters. It's a long-term event. It's one of the few disasters that people can **mitigate** if they pay attention to early warnings. For example, if people listen to an instruction to stop watering lawns, they'll save water. More restrictions might not be necessary.

CLIMATE FACTORS

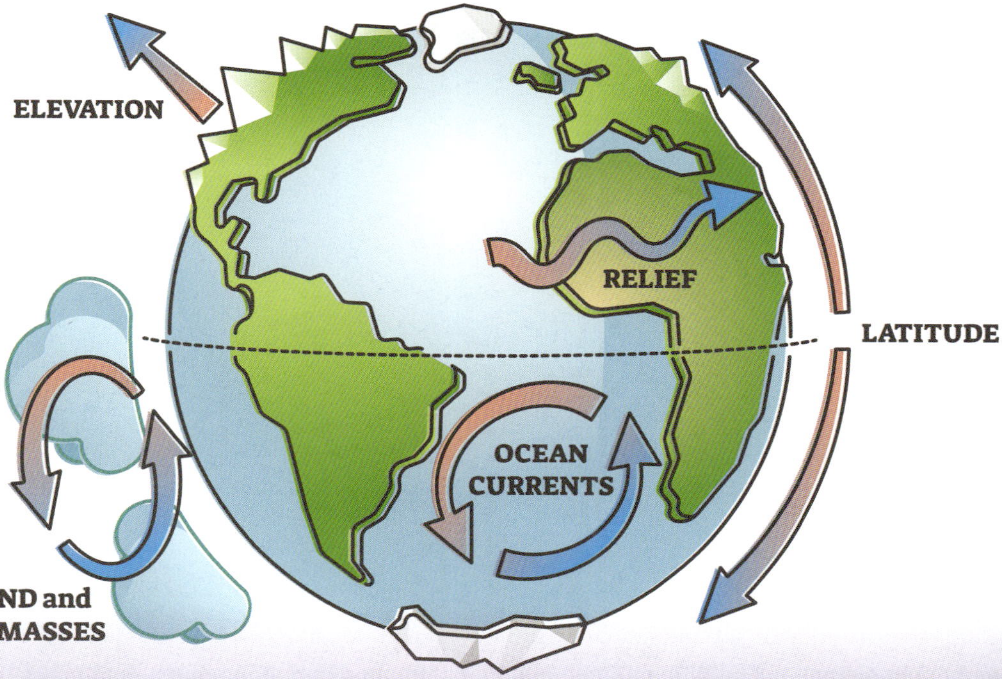

Scientists study a lot of different weather data to try to predict droughts. They also consider human activity. They might study an area's history to see if there have been previous droughts.

VERIFIED

The National Oceanic and Atmospheric Administration (NOAA) is part of the U.S. Department of Commerce. NOAA is the agency that monitors weather and climate. It issues official warnings and watches about weather emergencies. You can find out more at **www.noaa.gov**.

WHAT'S THE WORST THAT COULD HAPPEN?

Drought is a threat to **agriculture** around the world. If there's no rain, crops die. Without plant roots to hold it, rich topsoil can blow away. If farms can't grow crops, food can become very expensive. It might even become so scarce that people go hungry. This is called famine.

DISASTER RELIEF FOR FARMERS

The United States Department of Agriculture (USDA) has programs to help farmers recover from natural disasters. It provides aid when crops or livestock are lost. It also helps with water conservation projects.

VERIFIED

Oxfam America is part of a global organization dedicated to ending world hunger. It provides emergency aid to places experiencing famine. It also campaigns for human rights. You can learn more about it at **www.oxfamamerica.org**.

When crops fail, farmers can't pay their bills. They can't afford to buy seeds and equipment for the coming year. They might even lose their farm.

Famine is more likely to happen in less economically developed countries (LEDCs). These are places where farming and fishing are important industries. People who live there tend to be poor.

WHO STUDIES THIS STUFF?

The type of weather experienced in a place over time is called its climate. Scientists who study climate are called climatologists. Scientists who study what the climate was like long ago are called paleoclimatologists. Scientists who study weather are called meteorologists.

That's a lot of "ologists"! Droughts are complex, long-term events. It takes a lot of specialists to figure out if one is likely to happen. Scientists study things like tree rings and fossils to see if an area has experienced droughts in the past.

WOULD YOU LIKE TO BE A CLIMATOLOGIST?

Climatology is a diverse career path. Some climatologists work in the field. They study the environment up close. Others work in labs or classrooms. Climatologists need to have a passion for science. Most entry-level jobs require a bachelor's degree. Some climatologists study until they achieve a Ph.D.

THERE ARE SEVERAL COMMON CLIMATES ON EARTH:

Tropical	Located near the equator; always hot and humid
Dry	Desert; receives very little precipitation
Temperate	Warm summers and mild winters
Continental	Cool summers and very cold winters
Polar	Located at the poles; always cold

Cloud seeding is the process of placing a substance such as dry ice into clouds. This helps the water vapor in the clouds condense into rain.

HOW DO YOU MEASURE A DROUGHT?

The Palmer Drought Severity Index (PDSI) is used to classify droughts. The scale ranges from -10 (dry) to +10 (wet). The PDSI is used to measure long-term drought. It was developed in 1965 by meteorologist Wayne Palmer.

The North American Drought Monitor (NADM) is a project that studies droughts in the United States, Canada, and Mexico. The three countries work together because they share a single landmass. Drought doesn't stop at national borders. It affects farming and water supplies on both sides.

NORTH AMERICAN DROUGHT MONITOR CATEGORIES:

D0	Abnormally dry
D1	Moderate drought
D2	Severe drought
D3	Extreme drought
D4	Exceptional drought

VERIFIED

The National Integrated Drought Information System (NIDIS) publishes weekly maps and **statistics** about drought in the United States. You can find out more at **www.drought.gov**.

Scientists use many different tools to study droughts. Some are high-tech satellites and computers. Others are as simple as measuring rods inserted into bodies of water.

THE DUST BOWL

The Dust Bowl refers to an area of the Great Plains. It's the site of the worst drought in U.S. history. In the 1930s, the area was affected by high temperatures and little rain. Farmers plowed up the native prairie grass to plant wheat. Because there were no tough grass roots to hold the soil in place, dust storms blew away tons of dirt. Some of the dirt carried by storms carried disease.

This drought happened during the Great Depression. Many farmers lost their land after crops failed. There were few other jobs available. It would take about a decade for the land to recover.

BISON AND THE GREAT PLAINS

The natural habitat of the Great Plains had balance. Bison roamed freely, grazing on prairie grass. They never stayed in one place for long, so the land recovered quickly. However, European settlers plowed the earth to plant new crops. They introduced farm-raised cattle that stayed in one place and ate all the grass. Over time, this made the land more **susceptible** to drought.

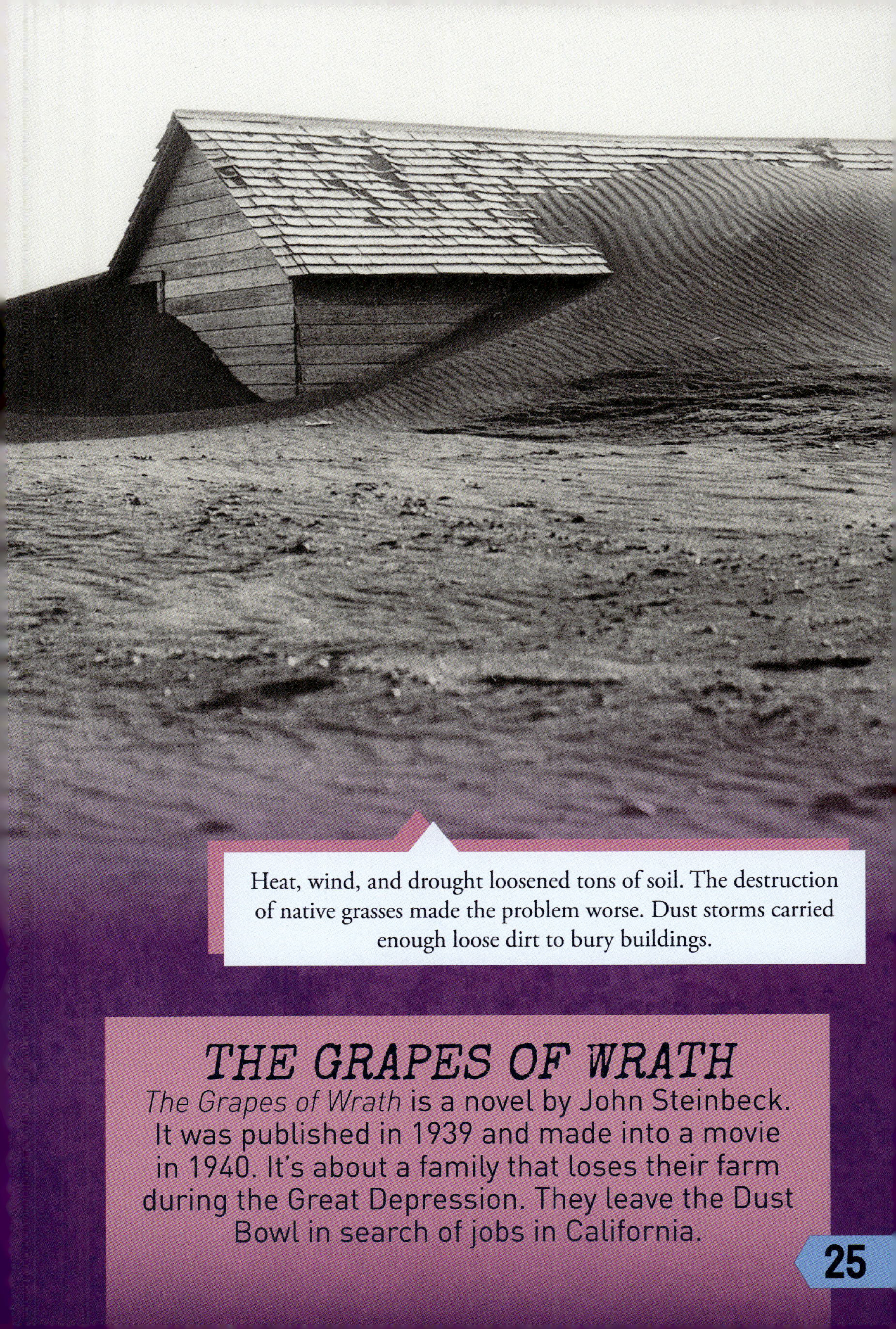

Heat, wind, and drought loosened tons of soil. The destruction of native grasses made the problem worse. Dust storms carried enough loose dirt to bury buildings.

THE GRAPES OF WRATH

The Grapes of Wrath is a novel by John Steinbeck. It was published in 1939 and made into a movie in 1940. It's about a family that loses their farm during the Great Depression. They leave the Dust Bowl in search of jobs in California.

WE ARE THE WORLD

Beginning around 1983, there was terrible drought and famine in Ethiopia, a country in Africa. Today, researchers think about 1 million people died. Musicians from the United States and United Kingdom (UK) organized projects to raise money to help the people there.

The UK project was called Band Aid. The group released a song called *Do They Know It's Christmas?*

U.S. musicians formed a group called U.S.A. for Africa. Their song was called *We Are the World*. On Good Friday, 1985, radio stations around the world played it at the same time.

LIVE AID

Live Aid was a rock concert held to benefit famine relief in Ethiopia. Two main concerts were held in London and Philadelphia on July 13, 1985. The event was broadcast around the world. It raised more than $127 million.

Band Aid, U.S.A. for Africa, and Live Aid helped make people aware of the disaster in Ethiopia. Many people hadn't known about the drought and famine. The songs and concert raised millions of dollars to help.

BOB GELDOF

Bob Geldof is an Irish rock musician. He organized Band Aid and Live Aid. He wanted to help people in Ethiopia. He earned an honorary knighthood and a Nobel Peace Prize nomination.

THE WORST EVER

Droughts can change the way people live. Lack of water can kill people and crops. Land can be damaged so badly that people need to move away. Even the **economy** and government can be affected.

Scientists believe people migrated, or moved, from Africa thousands of years ago because of a series of droughts. There wasn't enough water to drink or grow crops. Droughts are also thought to be factors in the fall of the Mayan empire and the Egyptian pharaohs.

DESERTIFICATION

Desertification is the process by which land becomes desert. This can happen as a result of drought or human activity. If all the plants in an area die off, the topsoil blows away. Without plants, the cycle of condensation that makes rain is disrupted.

THE 5 DEADLIEST RECORDED DROUGHTS

WHERE	WHEN	DEATHS
China	1928	3,000,000
Bangladesh	1943	1,900,000
India	1942	1,500,000
India	1965	1,500,000
India	1900	1,250,000

China has experienced many droughts and famines throughout history. The worst was in 1928. There were 3,000,000 deaths recorded. Historians now think the death toll was much higher.

LIFE DURING A DROUGHT

During a drought, governments will often set rules about water use. This is to make sure there's enough for everyone. People and animals need to drink. Crops need to be irrigated.

There might be rules about when people can water their lawns. Public fountains might be shut off. People might be asked not to wash their cars or driveways with a hose.

If a drought becomes bad enough, the government might forbid watering lawns at all. Public pools might not open.

WILDLIFE DURING A DROUGHT

During a drought, there's less food and water for wildlife. Wild animals might come right into your yard in search of food. They can make a mess of your garden or trash cans. They might also have diseases or parasites that your pets can catch.

Lawn grass isn't native to places like California. A small lawn can use 22,000 gallons (83,280 L) of water a year. That might be half the total water use for the household. This is very wasteful.

BEING MINDFUL

Drought is never very far away. Pollution and climate change make drought more likely. At the time this book was written, about half the United States was experiencing some level of drought.

There are many things people can do to help. It's always good to be aware of how much water we're using. Homeowners should check their houses for leaky plumbing. They can install toilets and appliances that use less water.

Cars can be taken to a professional car wash that recycles the water it uses.

XERISCAPING

Xeriscaping is a type of landscaping for arid, or very dry, climates. It involves:

- Improving the soil to suit the local climate
- Selecting plants that don't need much water
- Watering efficiently
- Using mulch to protect the soil

Some places are replacing grass with plants better suited to the local climate. Drought-**tolerant** plants are colorful and attractive. They thrive on very little water.

IN THE NEWS

Climate change and global warming are terms you hear a lot. Sometimes they seem to mean the same thing.

Climate change can be a natural process. An ice age is an example of a natural climate change. An ice age takes thousands of years to complete.

Global warming is a part of climate change. Earth's temperature has risen and cooled at different times in the past. However, the current trend of global warming is connected to pollution.

GRETA THUNBERG

Greta Thunberg is a Swedish student and environmental activist. She became interested in the topic of climate change at the age of eight. She changed her own lifestyle to be more environmentally friendly. She's spoken before governments and the United Nations.

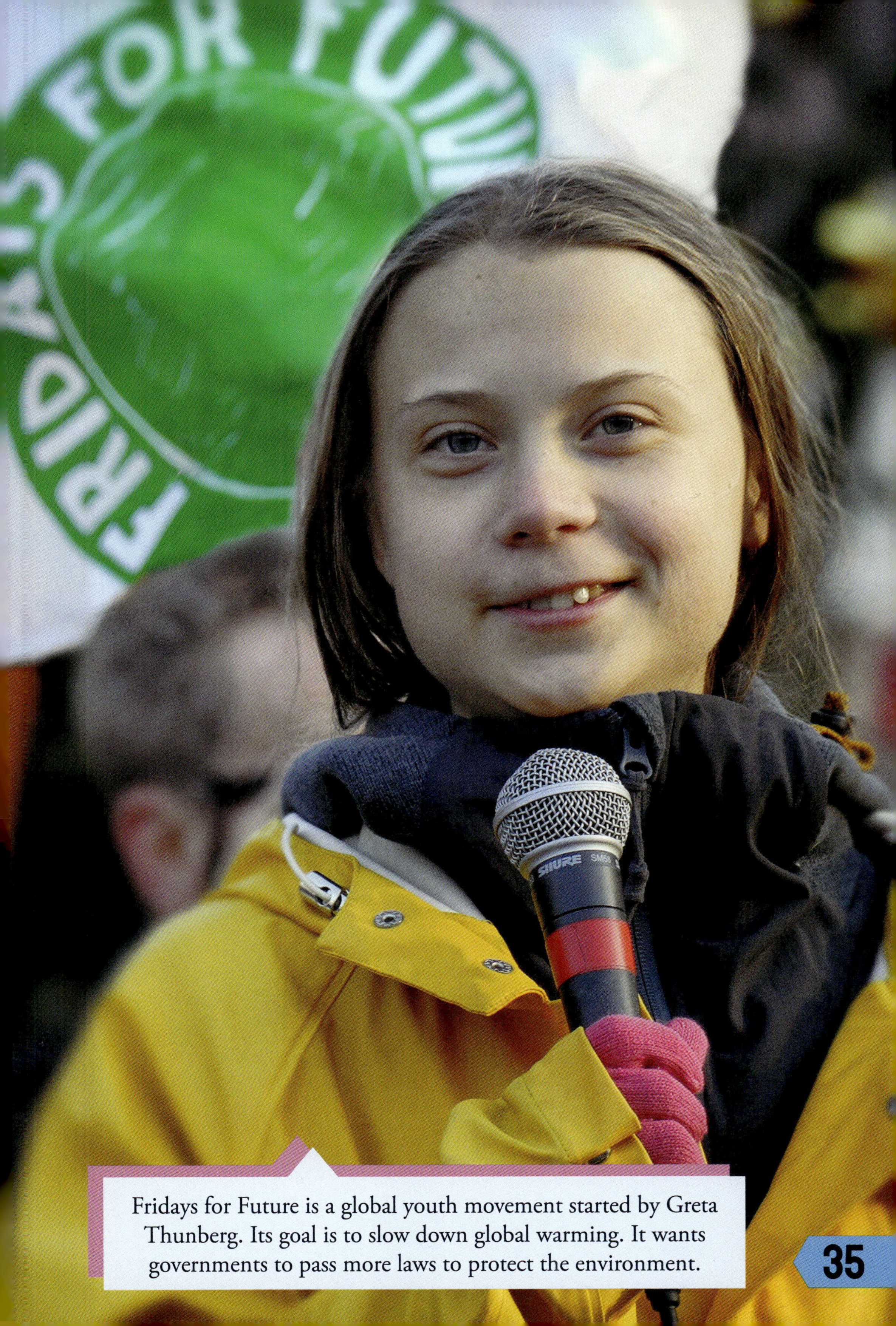

Fridays for Future is a global youth movement started by Greta Thunberg. Its goal is to slow down global warming. It wants governments to pass more laws to protect the environment.

TALK TO THE TREES

Dendrochronology is the study of the growth rings in trees. Each year of a tree's life, it grows another layer of wood. When the tree is cut down, each year's growth is visible as a ring. In years with plentiful water, the tree grows more. This shows as a thicker growth ring. During a drought, the tree grows less. This shows as a thinner ring.

This is a sample from a tree that's been cut down. The growth rings are clearly visible. They've been labeled by decade. This is valuable information about the tree and the place where it grew.

Scientists can study a tree's growth rings without cutting it down. They use a tool called an increment borer to take a long, skinny sample of the tree trunk. Then they count the growth rings on the sample.

Scientists study tree growth rings to figure out if an area has had droughts before. They can tell how long ago the droughts were and how long they lasted.

THE VIEW FROM ABOVE

Did you know that scientists can study droughts from space? Satellites are used to track greenhouse gases. In a recent project, **NASA** scientists made predictions about drought patterns for the rest of the century.

They used a thousand years of tree growth ring records to show past drought patterns. This information was entered into a supercomputer. The computer made a series of models based on how different levels of pollution affected global warming. The higher the levels of pollution, the worse the drought will be.

GREENHOUSE GASES

Methane and carbon dioxide are greenhouse gases. They occur naturally. They help to keep Earth warm enough to sustain life. So why do people say greenhouse gases are bad? It's because human activity produces much more of these gases than nature. They trap too much heat in Earth's atmosphere. This is changing climate patterns all over the planet.

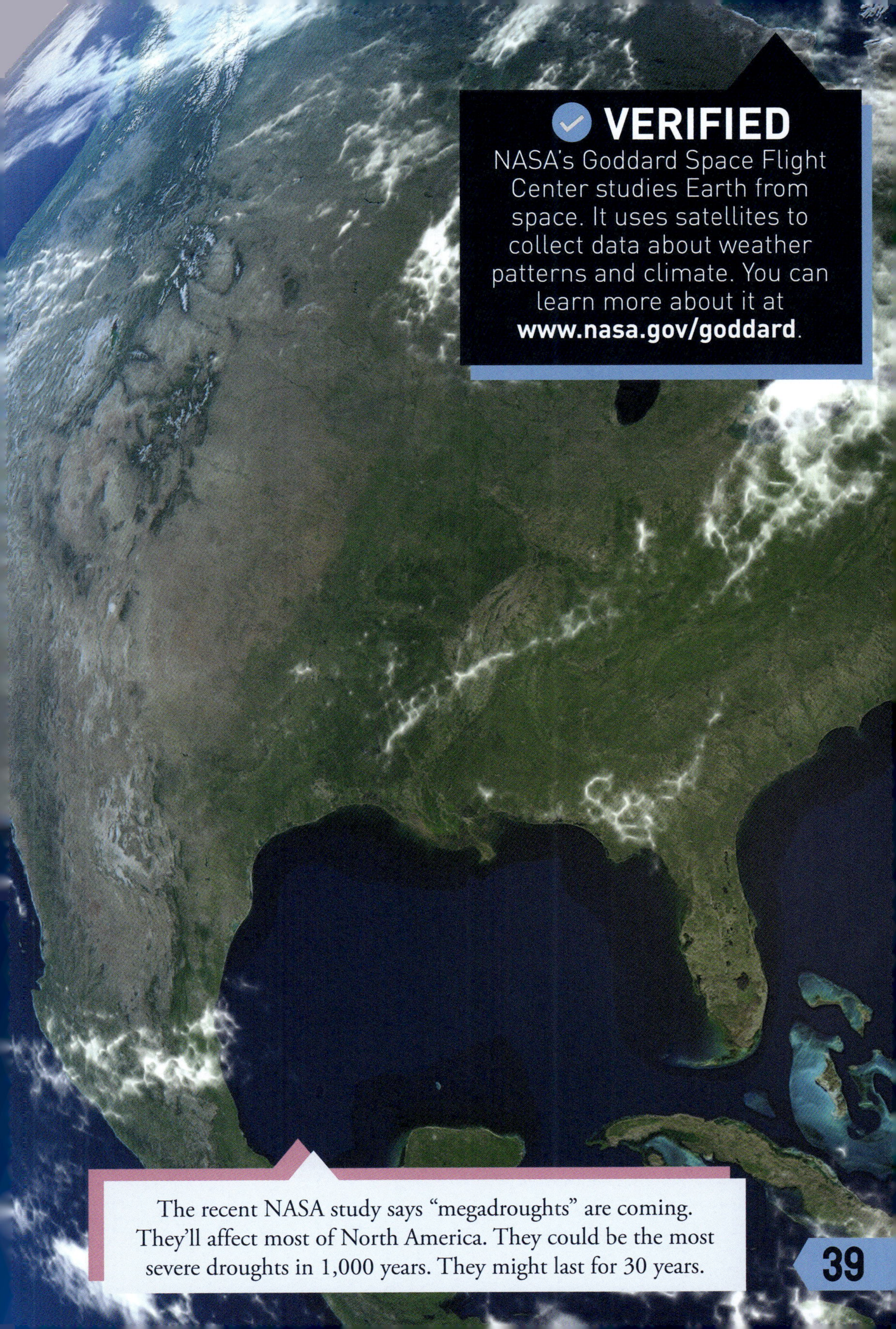

VERIFIED

NASA's Goddard Space Flight Center studies Earth from space. It uses satellites to collect data about weather patterns and climate. You can learn more about it at **www.nasa.gov/goddard**.

The recent NASA study says "megadroughts" are coming. They'll affect most of North America. They could be the most severe droughts in 1,000 years. They might last for 30 years.

WILDFIRE!

Wildfires are a natural occurrence. They can be started by lightning or even lava. Some types of plant life require fire as part of their life cycle.

WHEN IS FIRE GOOD FOR THE LAND?

A small natural fire, or one that's carefully controlled, is good for the land. Fire removes decayed plants from the forest floor. It clears out underbrush, allowing species that need more sunlight to thrive. Fire can also kill off **invasive species**.

Firefighters use chemicals to put out wildfires. The fire can still have an impact on water supplies. Ash can **contaminate** water supplies. Power outages might shut down water treatment plants.

Wildfires are being mentioned more often in the news. People have built homes in places where fire is part of the **ecosystem**. Drought conditions and human activity can make wildfires worse. Droughts provide plenty of dry fuel. Careless handling of matches or cigarettes can start a fire.

For over 75 years, Smokey Bear has educated the public about preventing forest fires. He's appeared on television. He's a familiar icon on signs throughout the park system.

DO PEOPLE CAUSE DROUGHTS?

A drought is a natural disaster. It has natural causes. Nothing people do can stop one from happening.

However, people's actions can affect a drought. Non-food crops that aren't suited for dry places waste a huge amount of water. Burning fossil fuels such as gasoline adds to global warming. As Earth's temperature rises, drought becomes more likely.

Recycling is another way to conserve water. It takes less water to recycle a resource than it does to create more.

WHAT SHOULD WE BE RECYCLING?

- Paper/clean cardboard
- Metal
- Glass
- Plastic

***Check your community's website for exact guidelines.**

VERIFIED

Greenpeace is an organization dedicated to protecting Earth. It lobbies for laws to protect the environment. It also organizes cleanup projects for polluted places. Find out more about it at **www.greenpeace.org**.

We all need to be mindful of our impact on the environment. However, large corporations do more damage than individuals. Do an internet search for the best and worst corporate polluters. Then decide whom you want to do business with.

WHAT DOES THE FUTURE HOLD?

There will probably always be cycles of drought on Earth. We can't prevent it. What we can do is change our behavior. We can recycle, landscape with the environment (not against it), and reduce our use of fossil fuels. These are all good steps.

A recent NASA study predicts a series of megadroughts for North America. There's a series of projections showing that a reduction in greenhouse gas emissions could make those droughts shorter and less severe.

Everyone's efforts can help. As an individual, make sure you're recycling. Turn off hoses and faucets when they're not in use. Remind your family members to run dishwashers and washing machines only when they're full.

We can't prevent droughts. If we all work together, we can protect our water supplies. We need to make sure there's enough for everyone.

The next time your parents are talking about a big purchase such as a car or appliance, remind them to look for environmentally friendly models. Do some research. Talk to your family about supporting companies that are kind to our planet.

When you're old enough to vote, support politicians who want to protect the environment.

We'll probably never live in a world without drought, but we can make sure there's enough water for everyone if we're careful.

GLOSSARY

adapt: To change your behavior so that it is easier to live in a particular place or situation.

agriculture: The science or occupation of farming.

conservation: The protection of animals, plants, and natural resources.

consumption: The use of something (such as fuel).

contaminate: To make something dangerous, dirty, or impure by adding something harmful or undesirable to it.

eco-friendly: Not harmful to the natural world.

economy: The process or system by which goods and services are produced, sold, and bought in a country or region.

ecosystem: Everything that exists in a particular environment.

environment: The natural world.

evaporate: To change from a liquid into a gas.

evolve: To change or develop slowly, often into a better, more complex, or more advanced state.

invasive species: A living thing that is not native to the place where found and tends to grow and spread easily, usually to the harm of native species and ecosystems.

mitigate: To make something less severe, harmful, or painful.

NASA: The National Aeronautics and Space Administration is an independent agency of the U.S. government responsible for the civil space program, aeronautics research, and space research.

statistic: A number that represents a piece of information (such as information about how often something is done, how common something is, etc.).

susceptible: Easily affected, influenced, or harmed by something.

tolerant: Able to allow or accept something that is harmful, unpleasant, etc.

INDEX